The ADULT ORGANIC Coloring Book

By Jacob E. Dander‡, Evan R. Darzi‡, and Neil K. Garg

COPYRIGHT

Copyright©2020 by Jacob E. Dander, Evan R. Darzi, and Neil K. Garg

All rights reserved. This book or any portion thereof may not be reproduced or used in any manner whatsoever without the express written permission of the publisher except for the use of brief quotations in a book review.

First Printing, 2020

ISBN: 979-8-5577-3104-1

Garg and Family Publishing Company
330 De Neve Drive
Los Angeles, CA 90024

Formatting and coordination by Daniel Caspi (Element26, Inc., element26.net)
Cover design and illustrations by Elena Kochetova

Chemical structures by Jacob E. Dander and Evan R. Darzi
Edited by Veronica Tona, Melissa Ramirez, and Elaina Garg

Email the authors: ngchem7@gmail.com
Visit Professor Neil Garg's Research Website: garg.chem.ucla.edu

Introduction by the Authors

"Why should kids have all the fun?"

In 2017, we released *The Organic Coloring Book* as a means to educate children about the importance of organic chemistry and the molecules that surround us. We were delighted to see that the book had a positive impact on children by helping to dispel the negative reputation of "chemicals" in modern society. However, we also appreciated the need to further share the importance of organic chemistry with a broader audience.

To meet this need, we decided to leverage the growing popularity of adult coloring books and hence concocted the *The Adult Organic Coloring Book*. Just as in *The Organic Coloring Book* and its successor, *Cheesy Goes to the Doctor*, the molecules discussed herein are organic. To a chemist, organic chemicals are molecules composed primarily of carbon and hydrogen atoms. They are all around us, including in drugs, foods, beverages, poisons, and more.

As you enjoy this coloring book, and accompany "Cheesy the Mouse" on a few adult-themed adventures, we hope you will come to appreciate that organic chemicals are everywhere and of paramount importance. When you look at the chemical structures, you will see that small changes in structure can profoundly change what molecules do. We encourage you to dive deeper into any topic from this book that interests you using the wealth of information you can find online.

Our society continues to make strides in improving scientific literacy, but we still have a long way to go. We hope *The Adult Organic Coloring Book* makes a contribution in this regard and serves to inspire the creation of related resources that positively impact society.

What else happens to ethanol in your body?

Some of the ethanol is converted into acetaldehyde. This molecule is partly responsible for hangovers. It is also responsible for the "alcohol flush reaction" (red cheeks!).

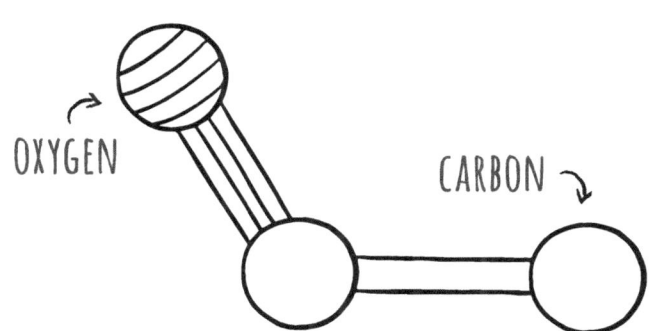

acetaldehyde

What's a chemical that can help build muscle?

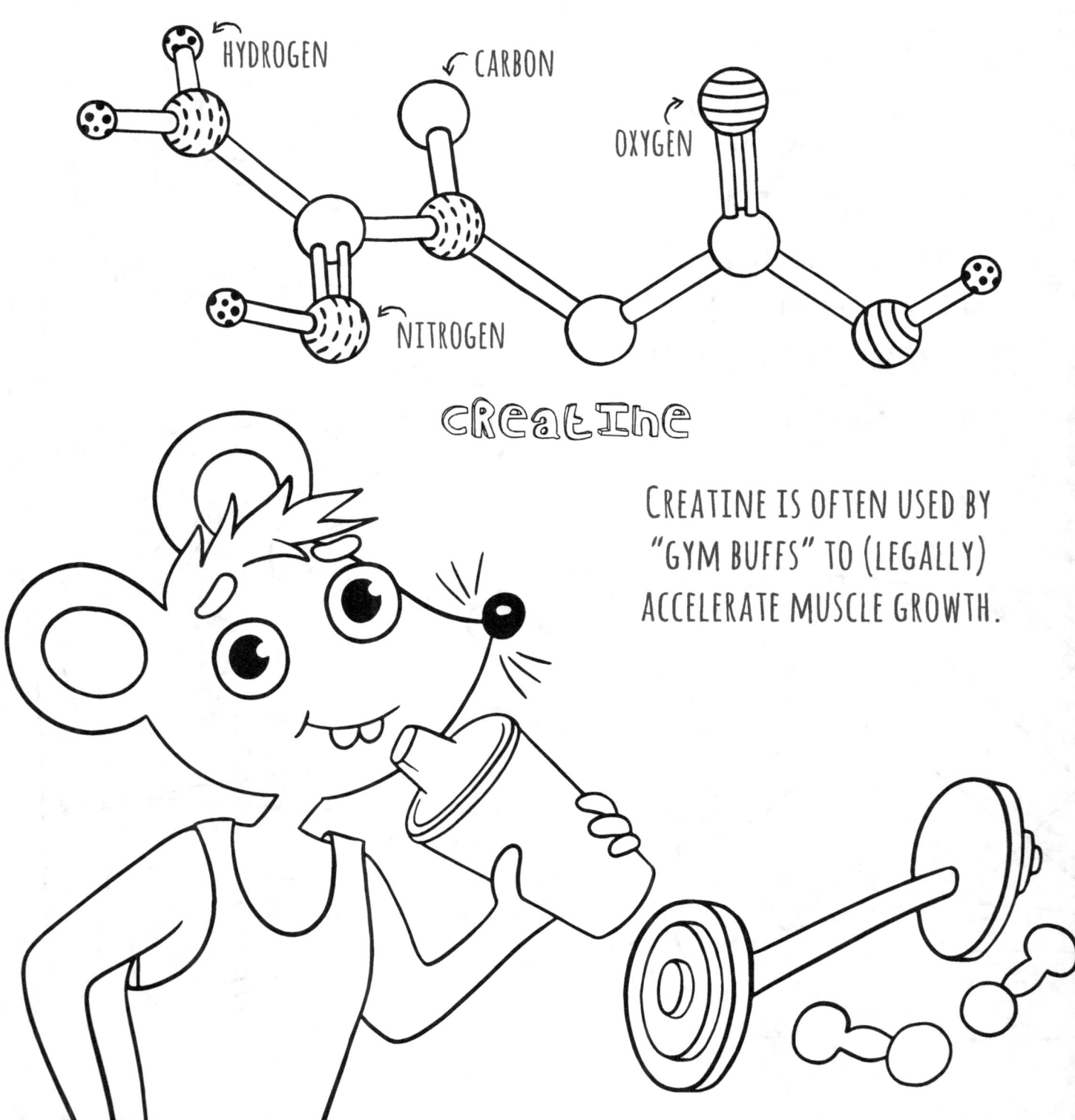

creatine

Creatine is often used by "gym buffs" to (legally) accelerate muscle growth.

How does creatine help to build muscle?

Creatine allows your body to produce more adenosine triphosphate (ATP)! ATP is an important energy "currency" for muscle growth.

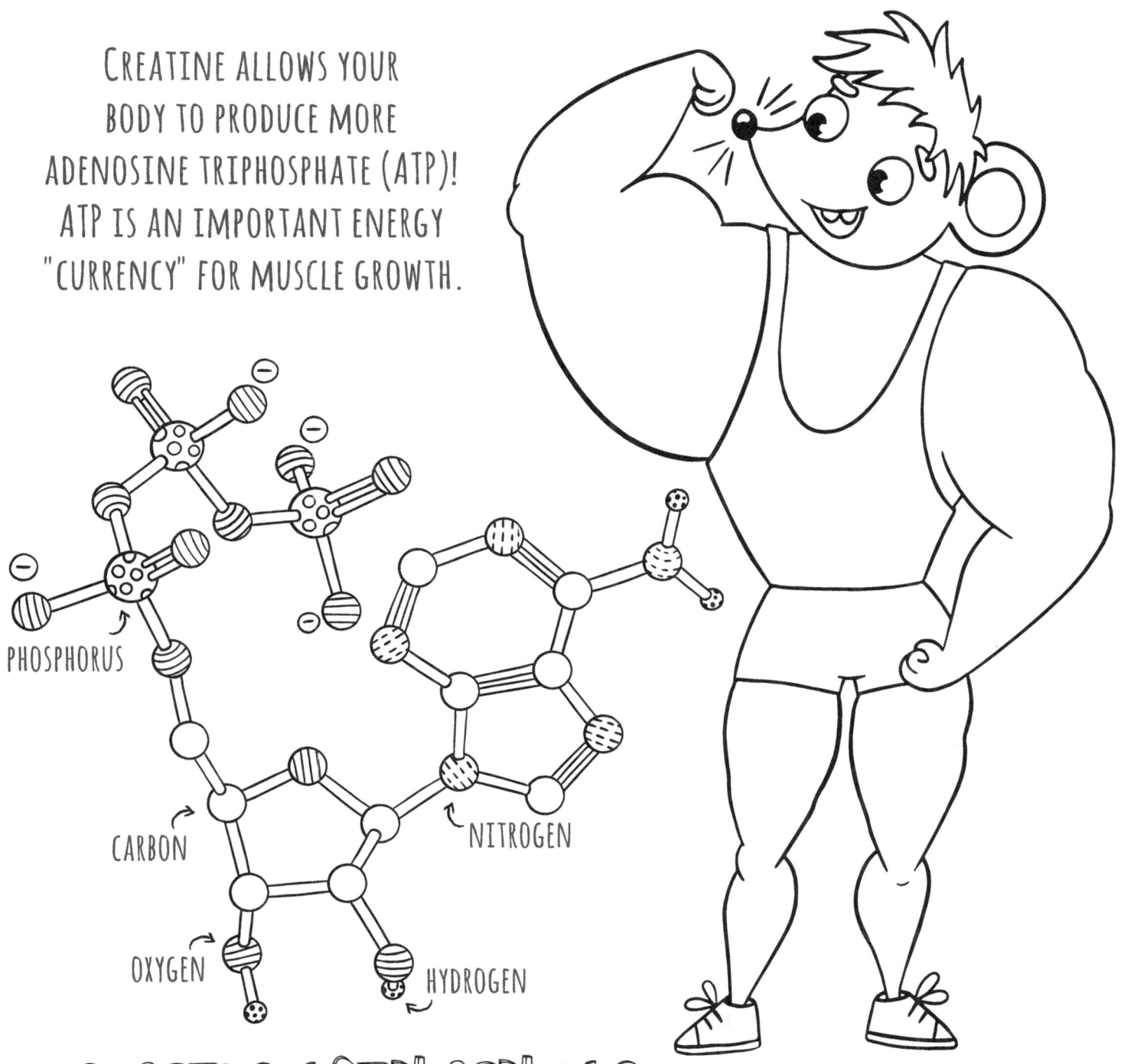

PHOSPHORUS

CARBON

NITROGEN

OXYGEN

HYDROGEN

adenosine triphosphate

LET'S TALK ABOUT FAT!

The chemicals we commonly associate with fat are called fatty acids. There are 2 major types: saturated and unsaturated.

STEARIC ACID

Saturated fatty acids are commonly found in meat and dairy products. WARNING: these fats can increase risk of heart disease.

LINOLEIC ACID

Unsaturated fatty acids have one or more double bonds, while saturated fatty acids do not.

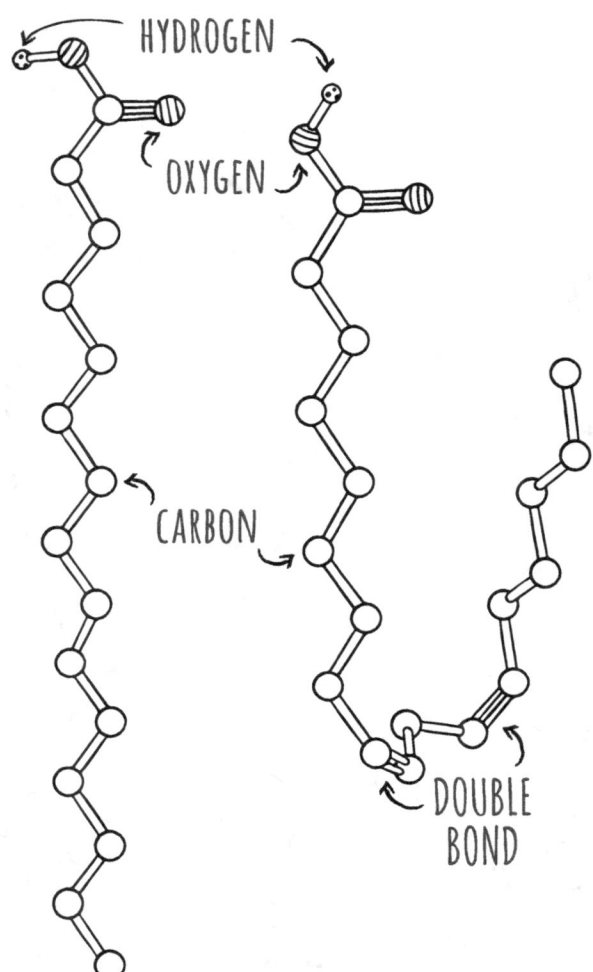

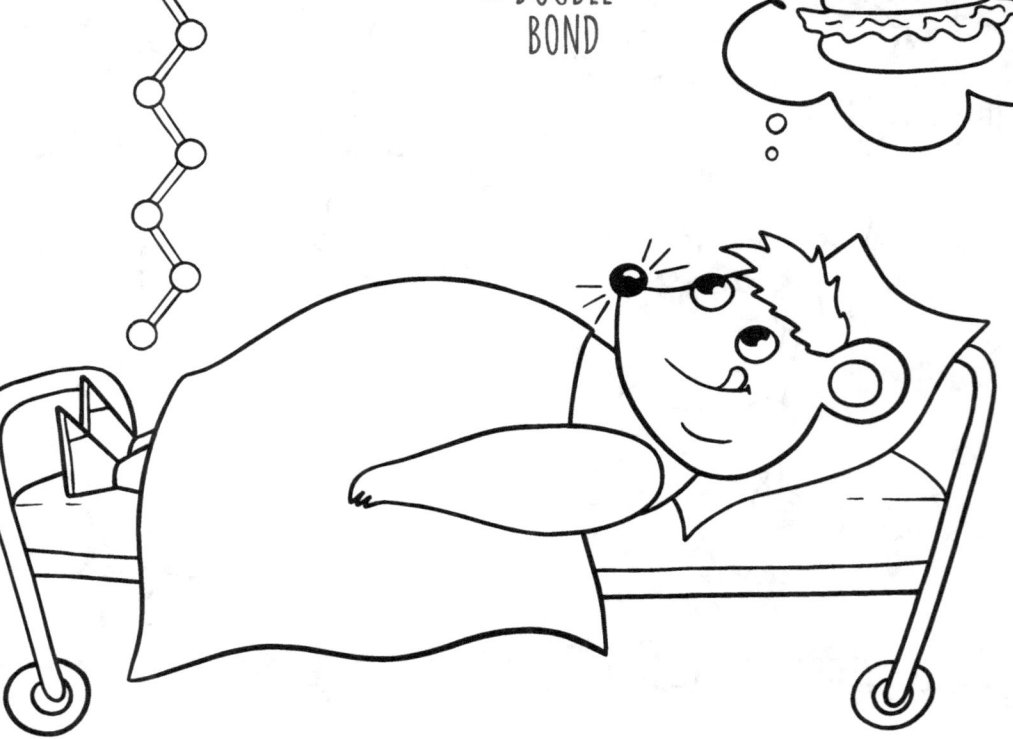

What About Cis and Trans Fats?

Unsaturated fatty acids come in two forms: cis or trans. Their small differences in shape can have big health consequences.

Elaidic Acid

Trans fats are often found in fried foods, like donuts and french fries. Similar to saturated fats, trans fats can increase your risk of heart disease.

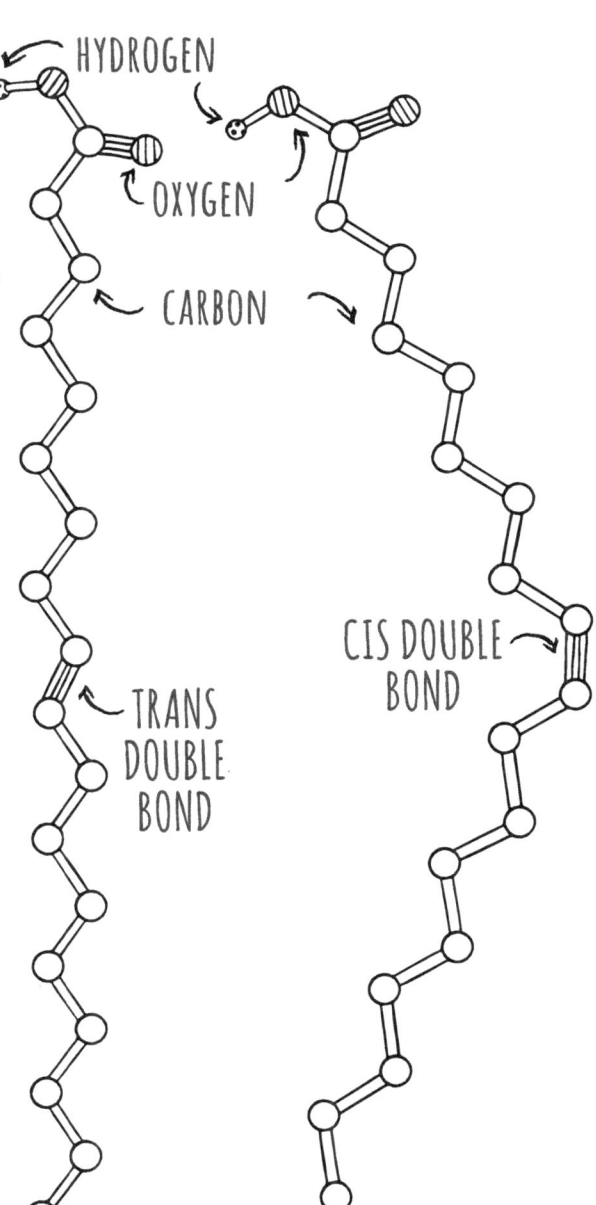

Hydrogen
Oxygen
Carbon
Trans double bond
Cis double bond

Oleic Acid

Cis fats can be found in fish and nuts. These are often considered to be the healthiest fats.

How Does Your Body Use Fat?

Fat has a bad reputation, but our bodies make good use of them!

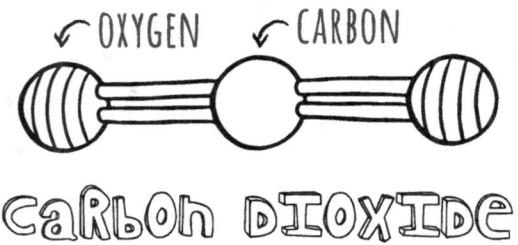

Carbon Dioxide

Some fats are metabolized to create energy. This releases carbon dioxide (CO_2) and generates ATP (ATP = energy).

Fats are also used to make other molecules.

The unsaturated fatty acid arachidonic acid is used to make Prostaglandin E2, which serves many roles in our bodies and is an important labor-inducing medicine used worldwide.

What's all the "rage" about testosterone?

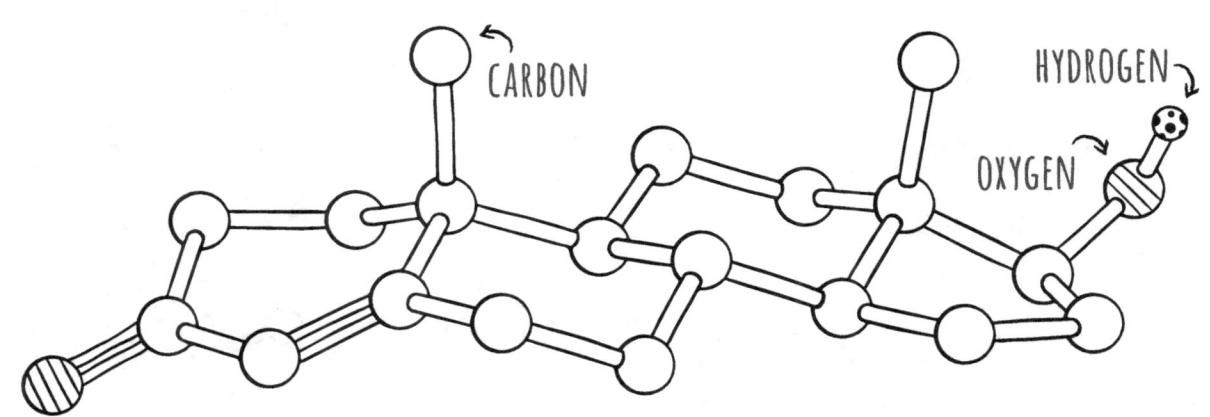

Testosterone

Testosterone is a hormone made in our bodies. It leads to muscle development and hair growth. Testosterone injections have been used illegally as performance enhancers in athletics.

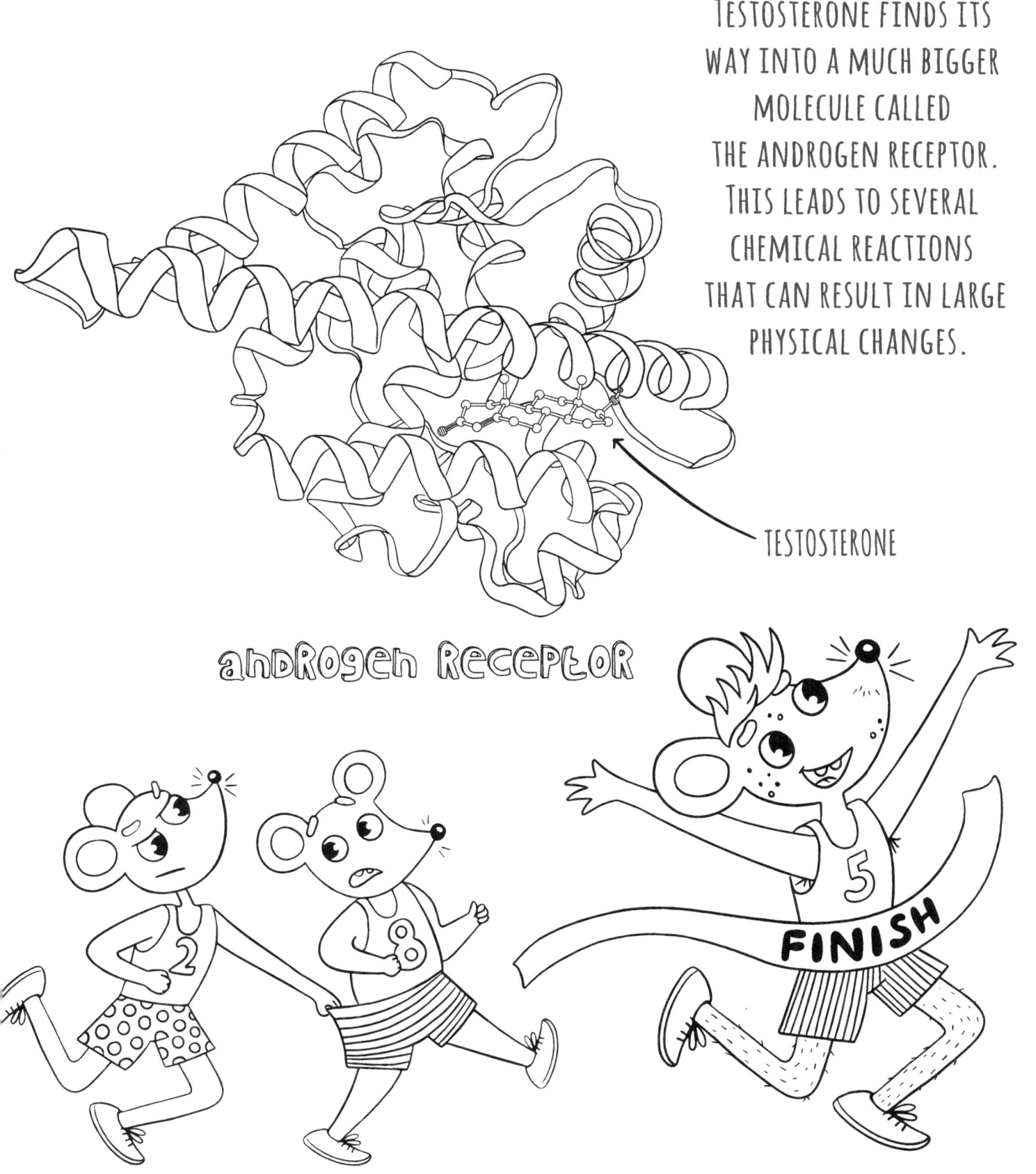

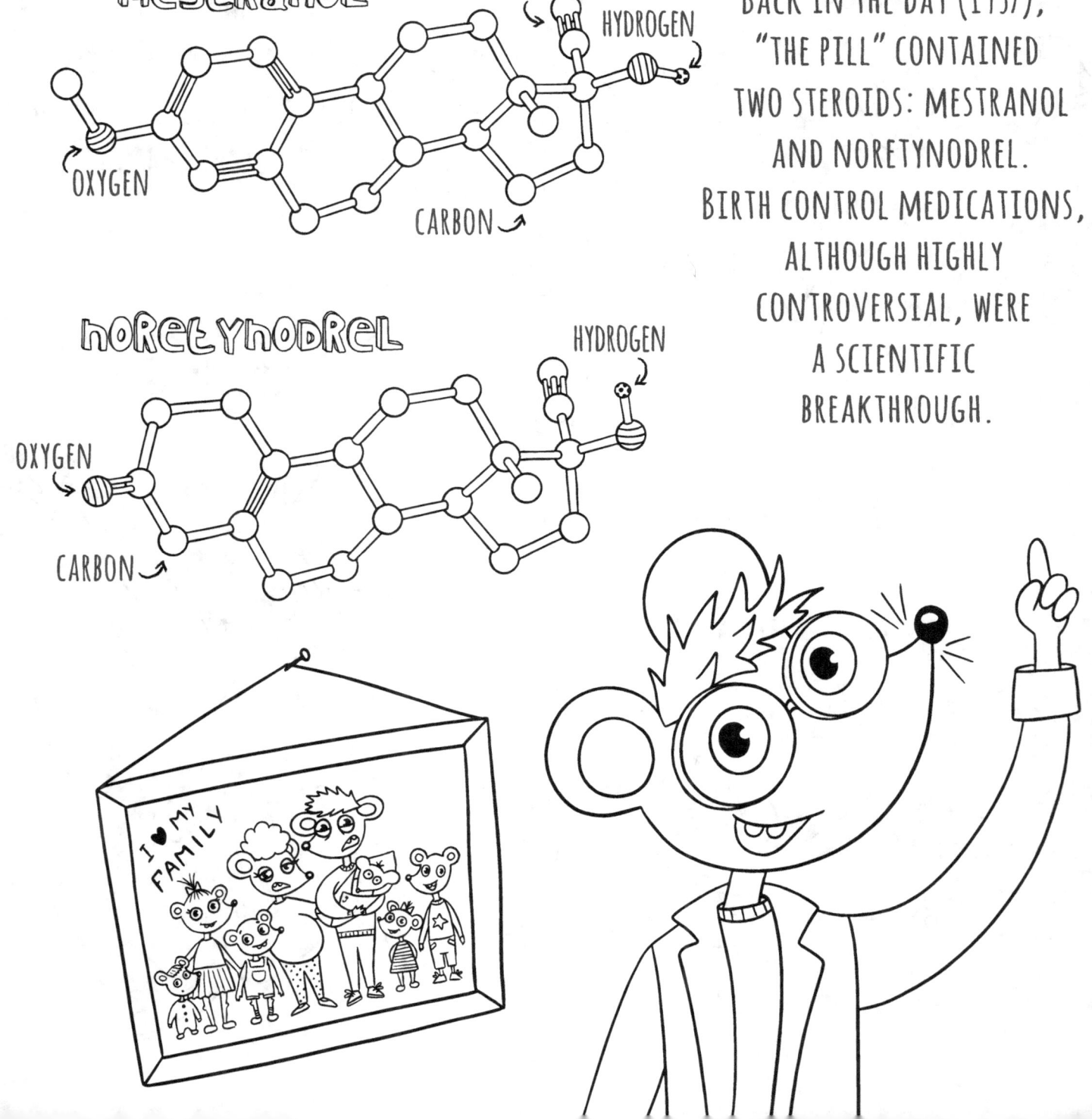

Are the same chemicals still used today?

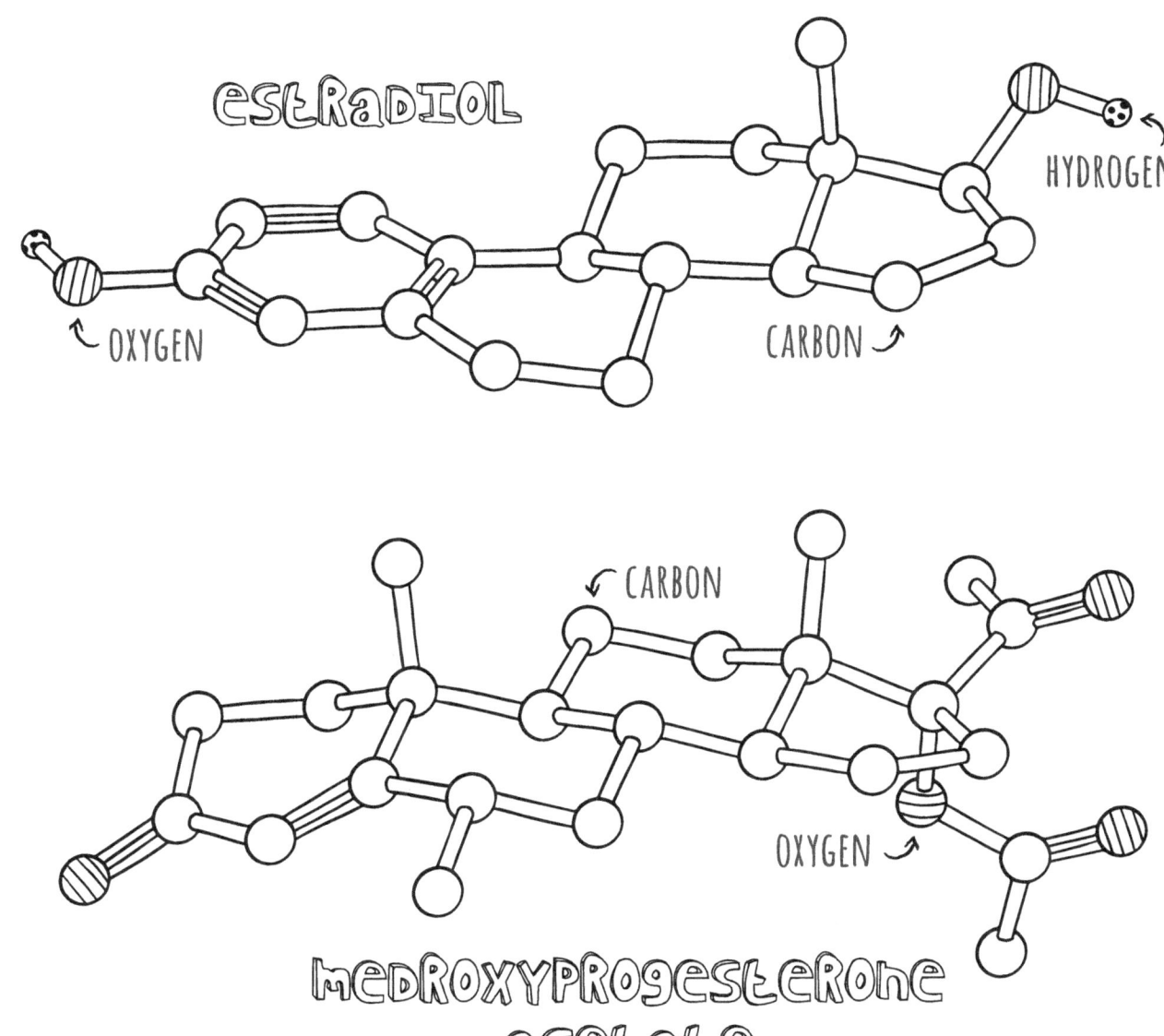

Birth control pills still contain one or more steroids, but different chemicals are generally used. Estradiol and medroxyprogesterone acetate are commonly used in combination and result in fewer side effects compared to the original "pill".

DID YOU KNOW THAT PSEUDOEPHEDRINE AND "METH" HAVE SIMILAR CHEMICAL STRUCTURES?

Pseudoephedrine is a common nasal decongestant found in Sudafed®, whereas "meth" is an addictive and dangerous drug. The only differences are a single oxygen atom and the placement of a hydrogen!

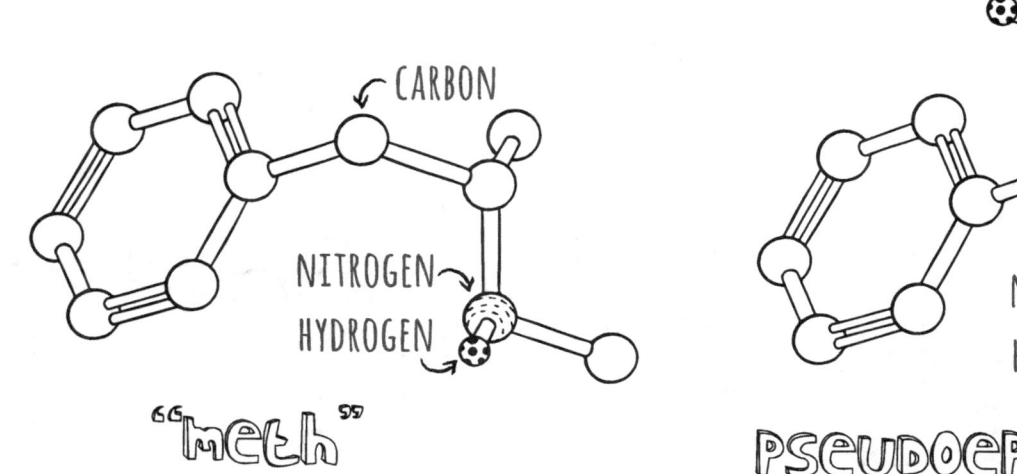
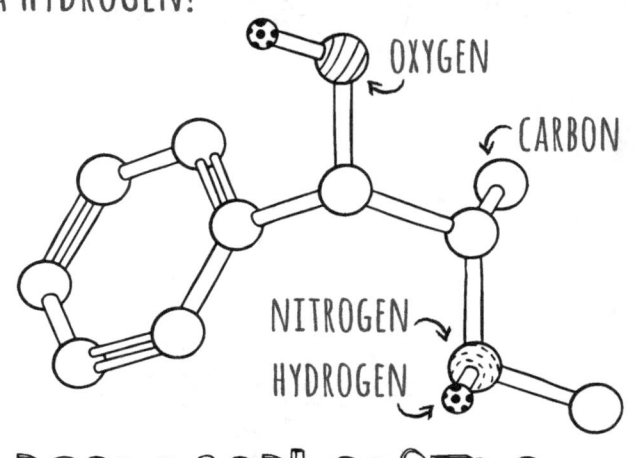

THE STRUCTURE OF "METH" IS ALSO REMARKABLY SIMILAR TO THAT OF ANOTHER MEDICINE.

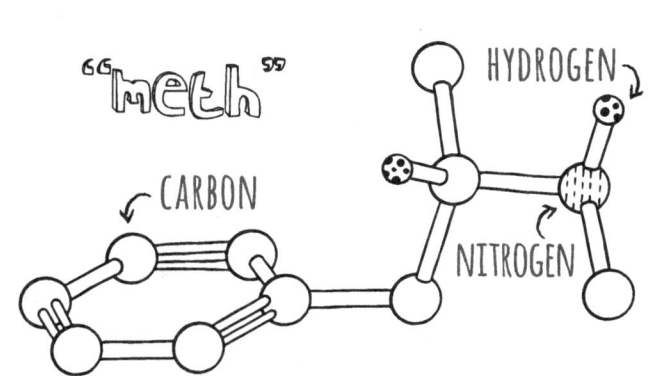

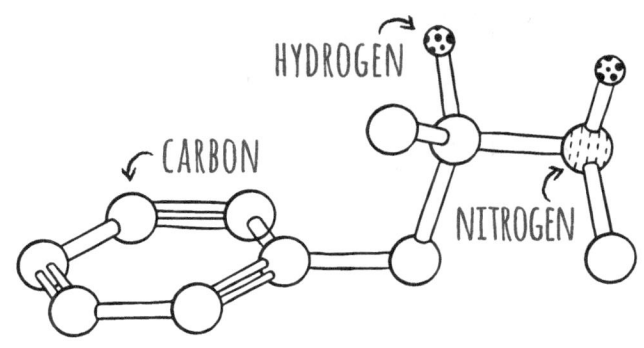

Levomethamphetamine is a nasal decongestant used in over-the-counter vapor inhalers. Its structure is identical to "meth," except some atoms are arranged differently in 3D space.

What is sugar and why is it sweet?

There are actually many different sugars, but sucrose (table sugar) is the most familiar. Sugars interact with taste receptors on your tongue, giving them a sweet flavor.

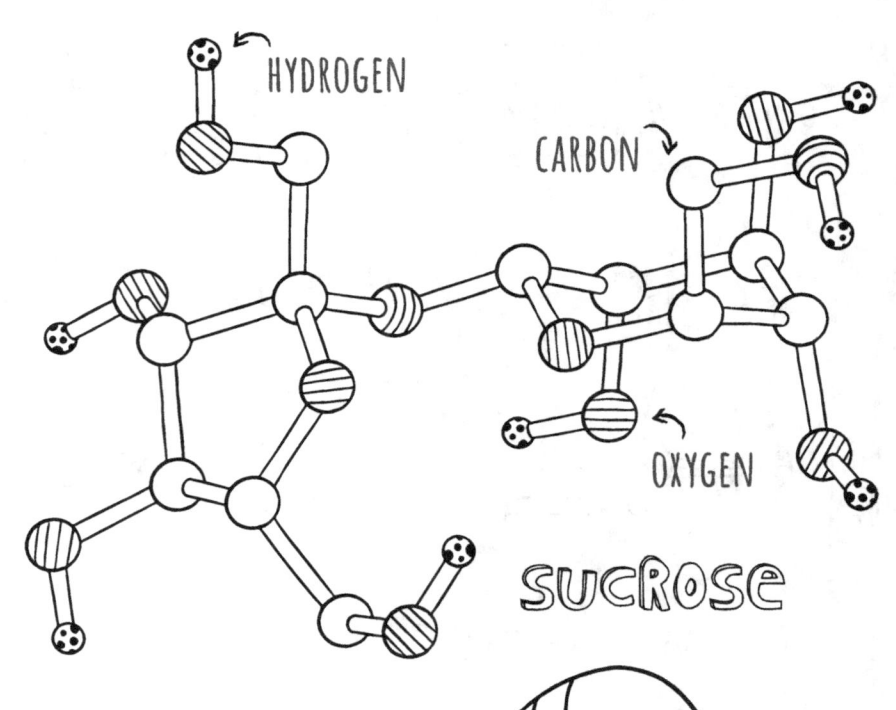

sucrose

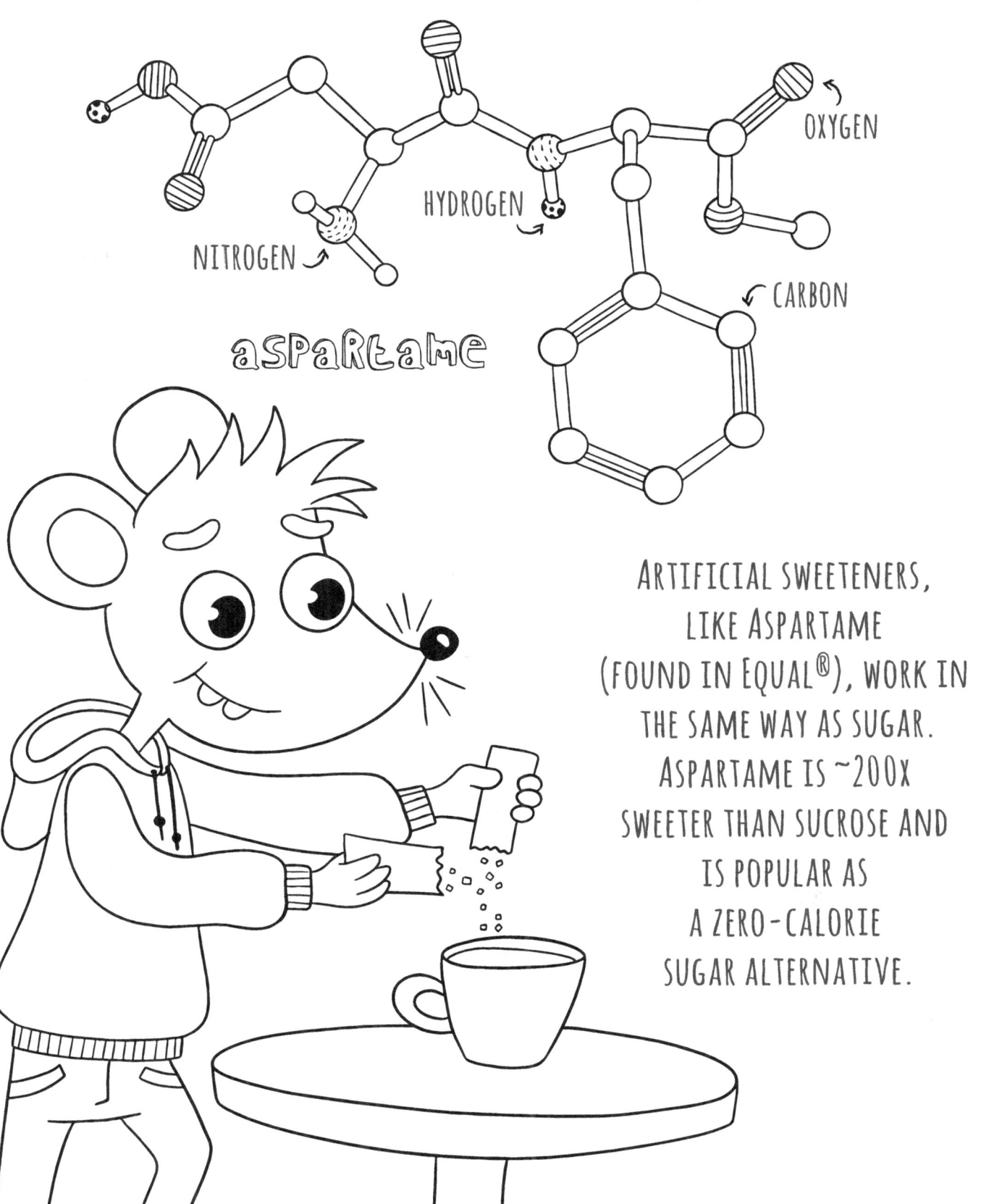

ARE CHEMICALS INVOLVED IN ALLERGIC REACTIONS?

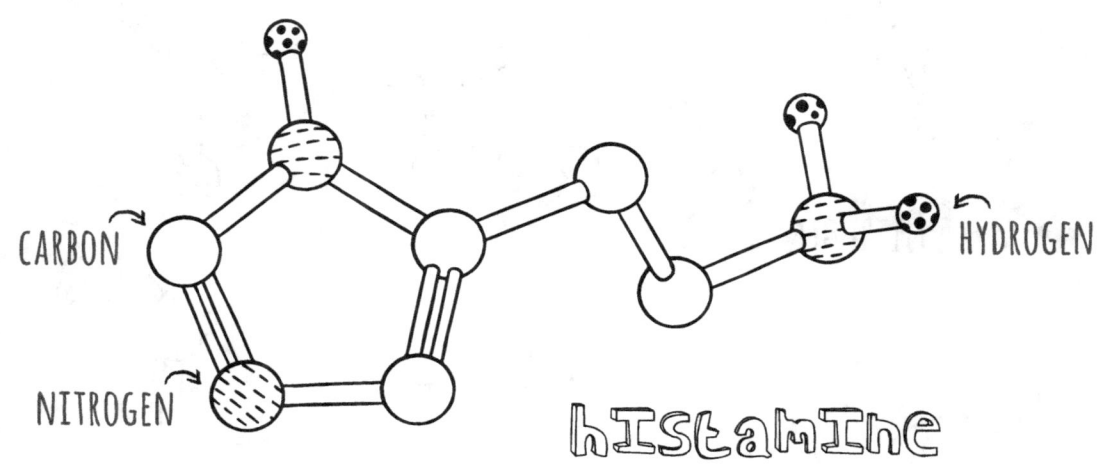

CARBON
HYDROGEN
NITROGEN
histamine

YES!
When we encounter an allergen, our bodies release histamine into our blood. This molecule then goes on to trigger our allergic responses like watery eyes, runny noses, and irritated skin.

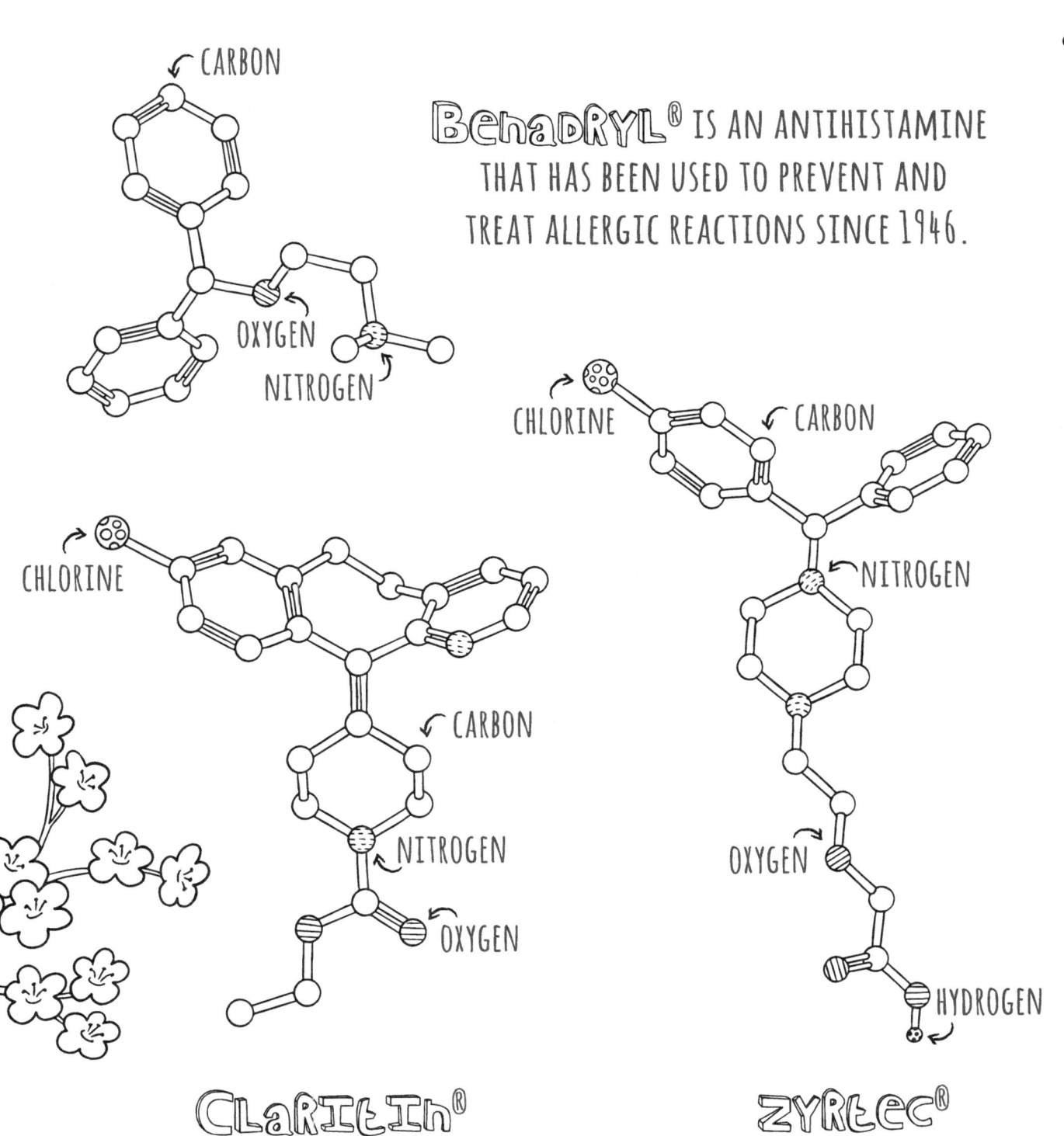

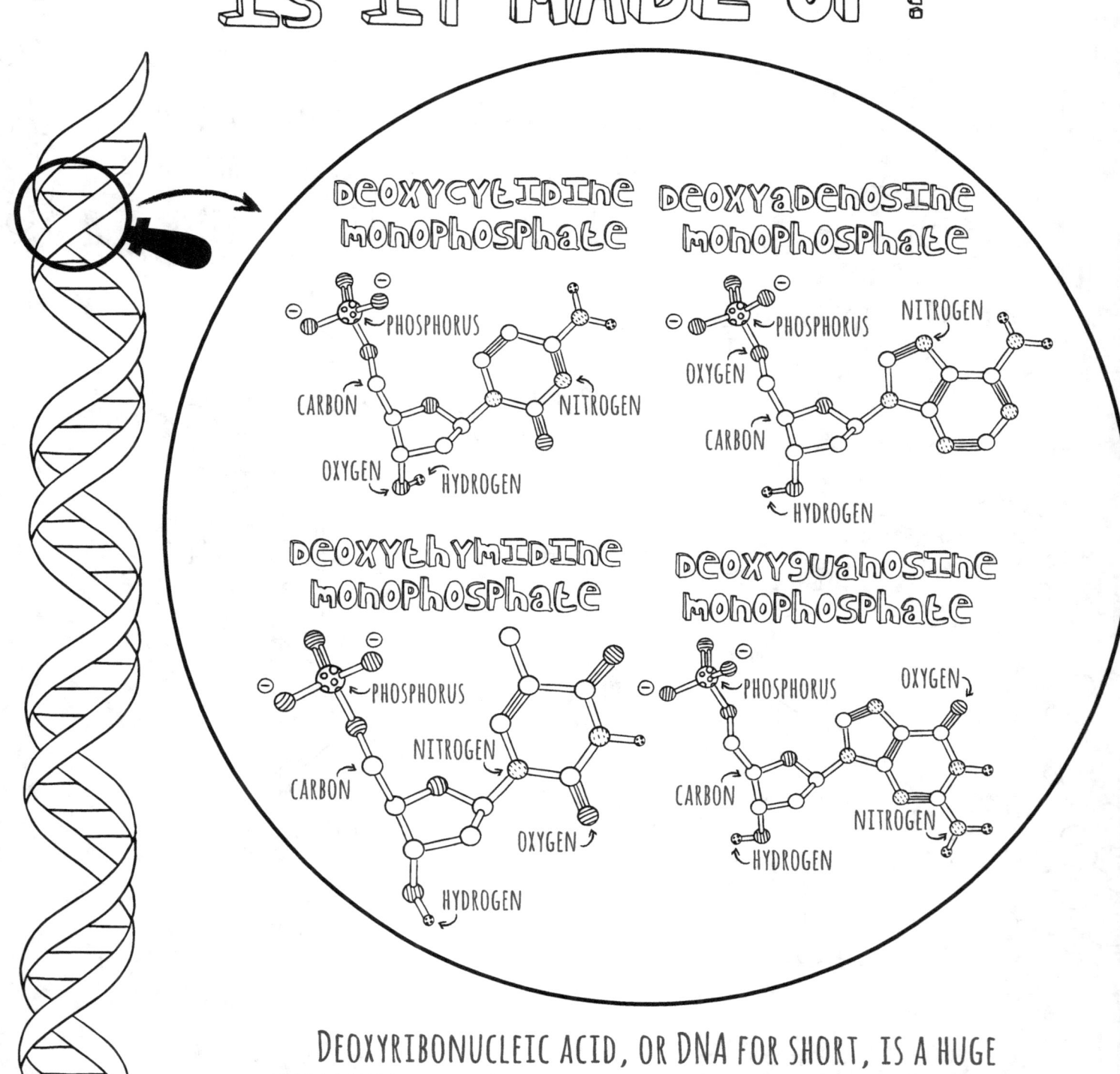

ARE NUCLEOTIDES USED FOR ANYTHING ELSE?

Many medicines resemble nucleotides or sections of DNA. The HIV/AIDS drug Retrovir®, commonly known as AZT (azidothymidine), is similar to the nucleotide thymidine monophosphate.

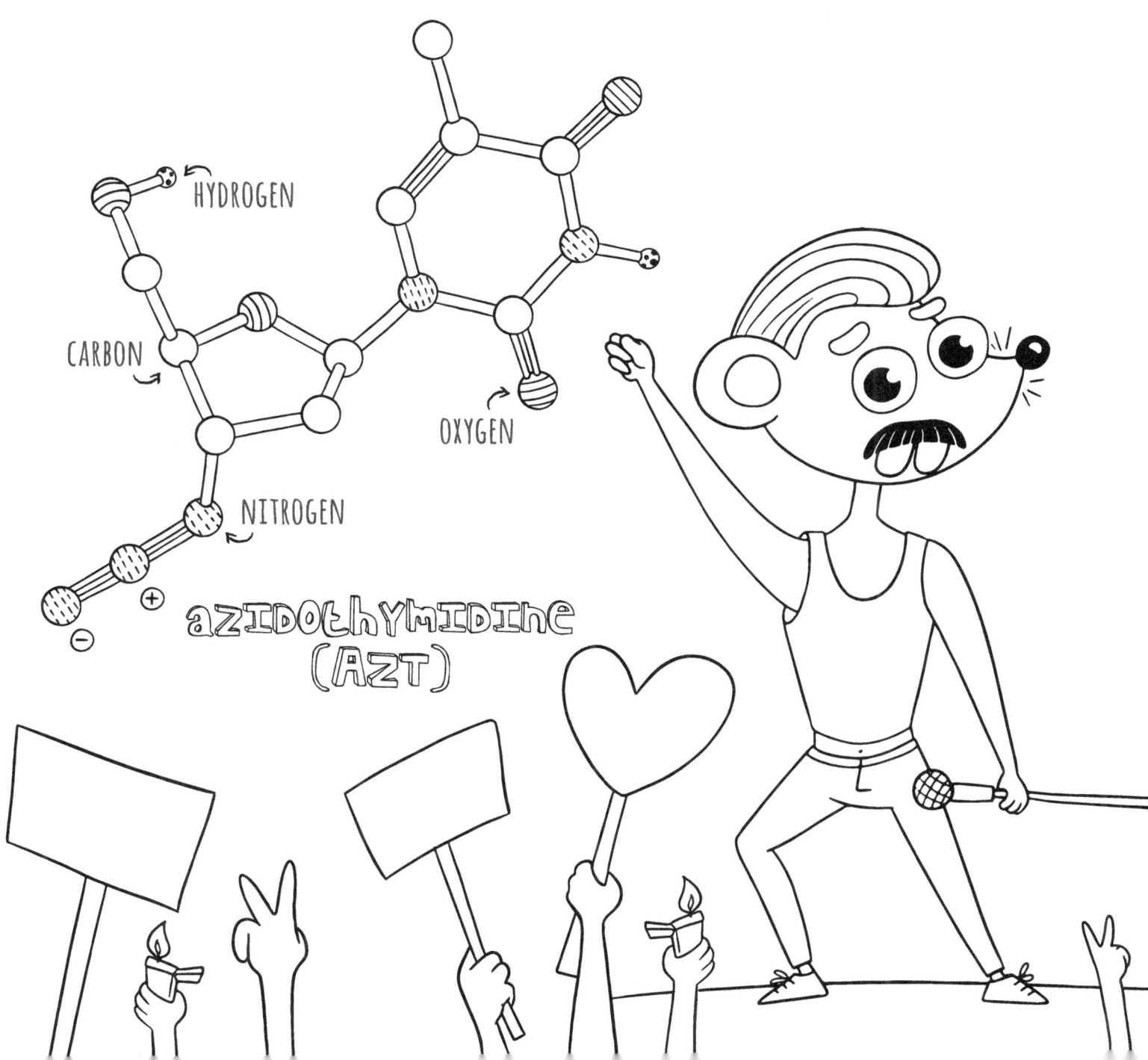

azidothymidine (AZT)

What chemicals are used to treat erectile dysfunction (ED)?

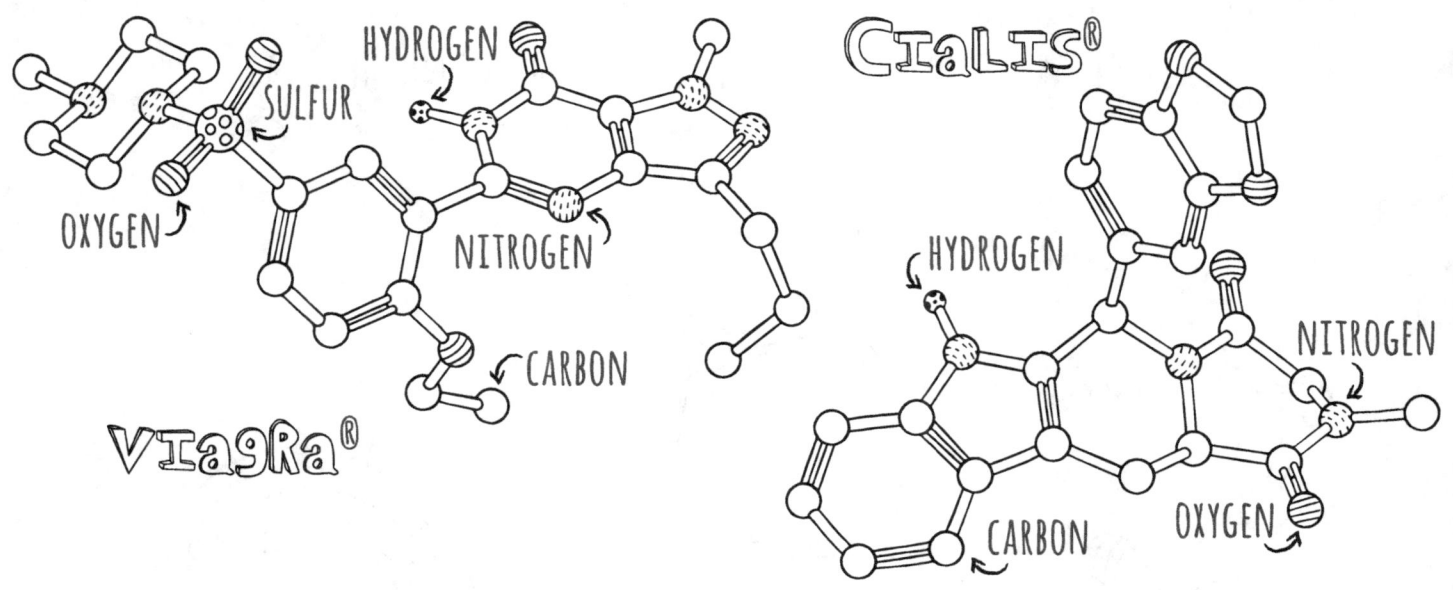

Viagra® (Sildenafil) and Cialis® (Tadalafil) are two of the most commonly prescribed ED medications.

HOW DO THEY WORK?

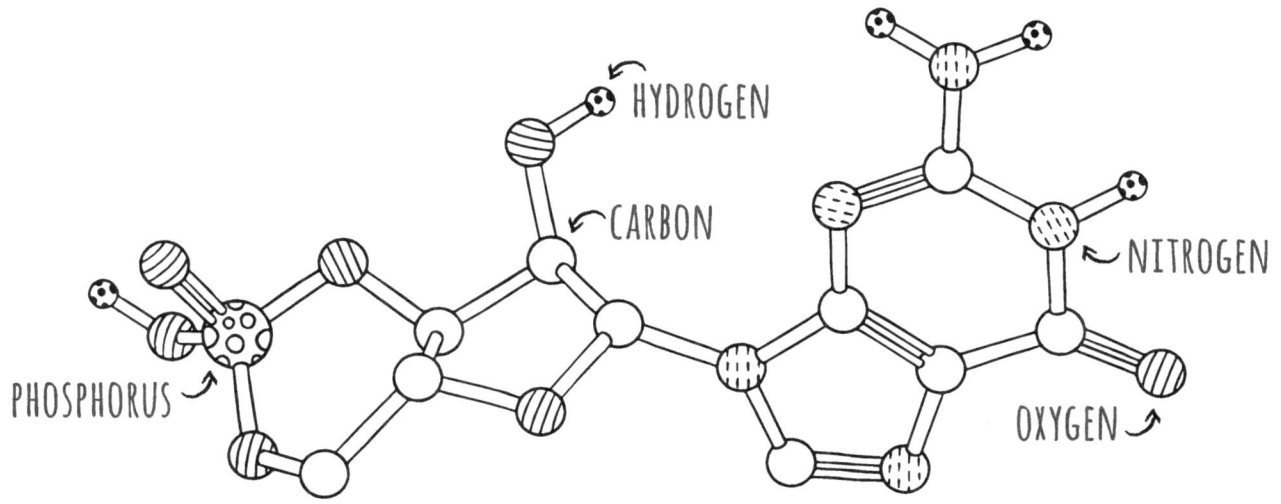

cyclic guanosine monophosphate (cGMP)

ED often results from poor blood circulation. Many ED medicines increase the amount of a chemical called cGMP in your blood, which leads to "enhanced" blood flow down low.

WHY DO SKUNKS STINK?

Skunks protect themselves by releasing a mixture of chemicals that smell really bad. The most offensive of them contain sulfur atoms attached to hydrogens, which are called thiols.

Do other animals use chemicals for defense?

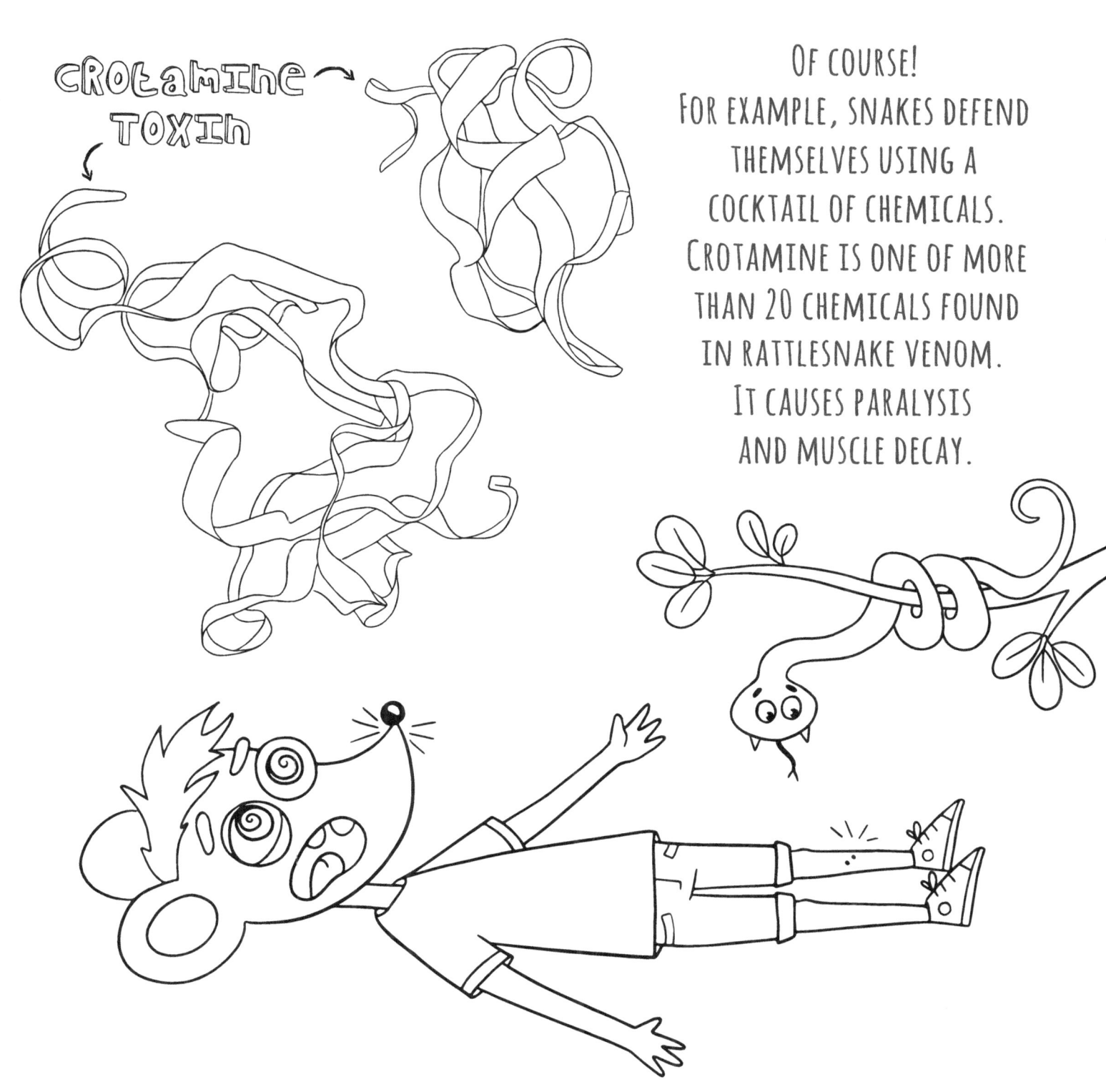

Crotamine toxin

Of course! For example, snakes defend themselves using a cocktail of chemicals. Crotamine is one of more than 20 chemicals found in rattlesnake venom. It causes paralysis and muscle decay.

www.ingramcontent.com/pod-product-compliance
Lightning Source LLC
Chambersburg PA
CBHW081710220526
45466CB00009B/2943